NOTEBOOK

CONTINUED FROM NOTEBOOK NO. _____   CONTINUED TO NOTEBOOK NO. _____

THIS NOTEBOOK BELONGS TO : _____

SIGNATURE : _____   DATE : _____

PERSONAL INFORMATION

PHONE NUMBER : _____   EMAIL ADRESS : _____

ADRESS : _____

CITY : _____   STATE : _____   ZIP : _____

NOTES

NOTEBOOK COMPLETED ON : _____   NUMBER OF PAGES FILLED IN : _____

TABLE OF CONTENTS

PAGE	SUBJET TITLE	DATE
1		
2		
3		
4		
5		
6		
7		
8		
9		
10		
11		
12		
13		
14		
15		
16		
17		
18		
19		
20		
21		
22		
23		
24		
25		
26		
27		
28		
29		
30		
31		
32		
33		
34		
35		
36		
37		
38		
39		
40		

BOOK NO._____

41		
42		
43		
44		
45		
46		
47		
48		
49		
50		
51		
52		
53		
54		
55		
56		
57		
58		
59		
60		
61		
62		
63		
64		
65		
66		
67		
68		
69		
70		
71		
72		
73		
74		
75		
76		
77		
78		
79		
80		
81		
82		
83		
84		
85		

BOOK NO._____

86		
87		
88		
89		
90		
91		
92		
93		
94		
95		
96		
97		
98		
99		
100		
101		
102		
103		
104		
105		
106		
107		
108		
109		
110		
111		
112		
113		
114		
115		
116		
117		
118		
119		
120		
121		
122		
123		
124		
125		
126		
127		
128		
129		
130		

BOOK NO._____

131	
132	
133	
134	
135	
136	
137	
138	
139	
140	
141	
142	
143	
144	
145	
146	
147	
148	
149	
150	
151	
152	
153	
154	
155	
156	
157	
158	
159	
160	
161	
162	
163	
164	
165	
166	
167	
168	
169	
170	
171	
172	
173	
174	
175	

BOOK NO._____

176		
177		
178		
179		
180		
181		
182		
183		
184		
185		
186		
187		
188		
189		
190		
191		
192		
193		
194		
195		
196		
197		
198		
199		
200		

BOOK NO._____

CONTINUED from page

CONTINUED on page

SIGNATURE:
WITNESSED BY:
DATE:
DATE:

BOOK NO.

3

CONTINUED from page

CONTINUED on page

SIGNATURE: _____ DATE: _____
WITNESSED BY: _____ DATE: _____

BOOK NO.

5

CONTINUED from page

CONTINUED on page

SIGNATURE: _____ DATE: _____
WITNESSED BY: _____ DATE: _____

BOOK NO.

6

CONTINUED from page

7

CONTINUED on page

SIGNATURE : _____ DATE : _____
WITNESSED BY : _____ DATE : _____

BOOK NO.

CONTINUED from page

CONTINUED on page

SIGNATURE: _____ DATE: _____
WITNESSED BY: _____ DATE: _____

BOOK NO.

8

CONTINUED from page

9

CONTINUED on page

SIGNATURE: _____ DATE: _____
WITNESSED BY: _____ DATE: _____

BOOK NO.

CONTINUED from page

10

CONTINUED on page

SIGNATURE: _____
WITNESSED BY: _____
DATE: _____
DATE: _____

BOOK NO.

CONTINUED from page

11

CONTINUED on page

SIGNATURE: _____ DATE: _____
WITNESSED BY: _____ DATE: _____

BOOK NO.

CONTINUED from page

12

CONTINUED on page

SIGNATURE: _____
WITNESSED BY: _____
DATE: _____
DATE: _____

BOOK NO.

CONTINUED from page

14

CONTINUED on page

SIGNATURE: _____ DATE: _____
WITNESSED BY: _____ DATE: _____

BOOK NO.

CONTINUED from page

15

CONTINUED on page

SIGNATURE: _____ DATE: _____
WITNESSED BY: _____ DATE: _____

BOOK NO.

CONTINUED from page

16

CONTINUED on page

SIGNATURE: _____ DATE: _____
WITNESSED BY: _____ DATE: _____

BOOK NO.

CONTINUED from page

17

CONTINUED on page

SIGNATURE: _____ DATE: _____
WITNESSED BY: _____ DATE: _____

BOOK NO.

CONTINUED from page

18

CONTINUED on page

SIGNATURE: _____ DATE: _____
WITNESSED BY: _____ DATE: _____

BOOK NO.

CONTINUED from page

CONTINUED on page

SIGNATURE: _____ DATE: _____
WITNESSED BY: _____ DATE: _____

BOOK NO.

CONTINUED from page

20

CONTINUED on page

SIGNATURE: _____ DATE: _____
WITNESSED BY: _____ DATE: _____

BOOK NO.

CONTINUED from page

CONTINUED on page

SIGNATURE: _____ DATE: _____
WITNESSED BY: _____ DATE: _____

BOOK NO.

CONTINUED from page

23

CONTINUED on page

SIGNATURE: _____ DATE: _____
WITNESSED BY: _____ DATE: _____   BOOK NO.

CONTINUED from page

25

CONTINUED on page

SIGNATURE: _____ DATE: _____
WITNESSED BY: _____ DATE: _____

BOOK NO.

CONTINUED from page

26

CONTINUED on page

SIGNATURE: _____ DATE: _____
WITNESSED BY: _____ DATE: _____

BOOK NO.

CONTINUED from page

CONTINUED on page

SIGNATURE: _____ DATE: _____
WITNESSED BY: _____ DATE: _____

BOOK NO.

27

CONTINUED from page

28

CONTINUED on page

SIGNATURE : _____ DATE : _____

WITNESSED BY : _____ DATE : _____

BOOK NO.

CONTINUED from page

29

CONTINUED on page

SIGNATURE: _____ DATE: _____
WITNESSED BY: _____ DATE: _____

BOOK NO.

CONTINUED from page

31

CONTINUED on page

SIGNATURE: _____ DATE: _____
WITNESSED BY: _____ DATE: _____

BOOK NO.

CONTINUED from page

32

CONTINUED on page

SIGNATURE: _____
WITNESSED BY: _____
DATE: _____
DATE: _____

BOOK NO.

CONTINUED from page

CONTINUED on page

SIGNATURE: _____ DATE: _____
WITNESSED BY: _____ DATE: _____

BOOK NO.

33

CONTINUED from page

34

CONTINUED on page

SIGNATURE: _____ DATE: _____

WITNESSED BY: _____ DATE: _____

BOOK NO.

CONTINUED from page

35

CONTINUED on page

SIGNATURE: _____ DATE: _____
WITNESSED BY: _____ DATE: _____

BOOK NO.

CONTINUED from page

CONTINUED on page

36

SIGNATURE: _____
WITNESSED BY: _____
DATE: _____
DATE: _____

BOOK NO.

CONTINUED from page

37

CONTINUED on page

SIGNATURE: _____ DATE: _____
WITNESSED BY: _____ DATE: _____

BOOK NO.

CONTINUED from page

CONTINUED on page

39

SIGNATURE: _____
WITNESSED BY: _____
DATE: _____
DATE: _____

BOOK NO.

CONTINUED from page

40

CONTINUED on page

SIGNATURE: _____
WITNESSED BY: _____
DATE: _____
DATE: _____

BOOK NO.

CONTINUED from page

41

CONTINUED on page

SIGNATURE: _____ DATE: _____
WITNESSED BY: _____ DATE: _____

BOOK NO.

CONTINUED from page

42

CONTINUED on page

SIGNATURE: _____ DATE: _____
WITNESSED BY: _____ DATE: _____

BOOK NO.

CONTINUED from page

CONTINUED on page

SIGNATURE: _____ DATE: _____

WITNESSED BY: _____ DATE: _____

BOOK NO.

43

CONTINUED from page

CONTINUED on page

SIGNATURE: _____ DATE: _____

WITNESSED BY: _____ DATE: _____

BOOK NO.

CONTINUED from page

45

CONTINUED on page

SIGNATURE: _____ DATE: _____
WITNESSED BY: _____ DATE: _____

BOOK NO.

CONTINUED from page

47

CONTINUED on page

SIGNATURE : _____ DATE : _____
WITNESSED BY : _____ DATE : _____

BOOK NO.

CONTINUED from page

48

CONTINUED on page

SIGNATURE: _____ DATE: _____
WITNESSED BY: _____ DATE: _____

BOOK NO.

CONTINUED from page

50

CONTINUED on page

SIGNATURE: _____ DATE: _____
WITNESSED BY: _____ DATE: _____

BOOK NO.

CONTINUED from page

51

CONTINUED on page

SIGNATURE: _____ DATE: _____
WITNESSED BY: _____ DATE: _____

BOOK NO.

CONTINUED from page

52

CONTINUED on page

SIGNATURE: _____
WITNESSED BY: _____
DATE: _____
DATE: _____

BOOK NO.

CONTINUED from page

53

CONTINUED on page

SIGNATURE: _____
WITNESSED BY: _____
DATE: _____
DATE: _____

BOOK NO.

CONTINUED from page

54

CONTINUED on page

SIGNATURE: _____   DATE: _____
WITNESSED BY: _____   DATE: _____

BOOK NO.

CONTINUED from page

55

CONTINUED on page

SIGNATURE: _____
WITNESSED BY: _____
DATE: _____
DATE: _____

BOOK NO.

CONTINUED from page

CONTINUED on page

SIGNATURE: _____
WITNESSED BY: _____
DATE: _____
DATE: _____

BOOK NO.

56

CONTINUED from page

57

CONTINUED on page

SIGNATURE: _____
WITNESSED BY: _____
DATE: _____
DATE: _____

BOOK NO.

CONTINUED from page

59

CONTINUED on page

SIGNATURE: _____ DATE: _____
WITNESSED BY: _____ DATE: _____

BOOK NO.

CONTINUED from page

60

CONTINUED on page

SIGNATURE: _____
WITNESSED BY: _____
DATE: _____
DATE: _____

BOOK NO.

CONTINUED from page

61

CONTINUED on page

SIGNATURE: _____
WITNESSED BY: _____
DATE: _____
DATE: _____

BOOK NO.

CONTINUED from page

62

CONTINUED on page

SIGNATURE: _____ DATE: _____
WITNESSED BY: _____ DATE: _____

BOOK NO.

CONTINUED from page

63

CONTINUED on page

SIGNATURE: _____ DATE: _____

WITNESSED BY: _____ DATE: _____

BOOK NO.

CONTINUED from page

64

CONTINUED on page

SIGNATURE: _____
WITNESSED BY: _____
DATE: _____
DATE: _____

BOOK NO.

CONTINUED from page

65

CONTINUED on page

SIGNATURE: _____
WITNESSED BY: _____
DATE: _____
DATE: _____

BOOK NO.

CONTINUED from page

66

CONTINUED on page

SIGNATURE:
WITNESSED BY:
DATE:
DATE:

BOOK NO.

CONTINUED from page

67

CONTINUED on page

SIGNATURE: _____ DATE: _____
WITNESSED BY: _____ DATE: _____

BOOK NO.

CONTINUED from page

68

CONTINUED on page

SIGNATURE: _____
WITNESSED BY: _____
DATE: _____
DATE: _____

BOOK NO.

CONTINUED from page

69

CONTINUED on page

SIGNATURE: _____
WITNESSED BY: _____
DATE: _____
DATE: _____

BOOK NO.

CONTINUED from page

71

CONTINUED on page

SIGNATURE: _____ DATE: _____
WITNESSED BY: _____ DATE: _____

BOOK NO.

CONTINUED from page

72

CONTINUED on page

SIGNATURE: _____ DATE: _____
WITNESSED BY: _____ DATE: _____

BOOK NO.

CONTINUED from page

73

CONTINUED on page

SIGNATURE: _____
WITNESSED BY: _____
DATE: _____
DATE: _____

BOOK NO.

CONTINUED from page

74

CONTINUED on page

SIGNATURE: _____ DATE: _____
WITNESSED BY: _____ DATE: _____

BOOK NO.

CONTINUED from page

75

CONTINUED on page

SIGNATURE: _____ DATE: _____
WITNESSED BY: _____ DATE: _____   BOOK NO.

CONTINUED from page

76

CONTINUED on page

SIGNATURE: _____
WITNESSED BY: _____
DATE: _____
DATE: _____

BOOK NO.

CONTINUED from page

CONTINUED on page

SIGNATURE: _____ DATE: _____

WITNESSED BY: _____ DATE: _____

BOOK NO.

77

CONTINUED from page

79

CONTINUED on page

SIGNATURE: _____ DATE: _____
WITNESSED BY: _____ DATE: _____

BOOK NO.

CONTINUED from page

80

CONTINUED on page

SIGNATURE: _____ DATE: _____
WITNESSED BY: _____ DATE: _____

BOOK NO.

CONTINUED from page

81

CONTINUED on page

SIGNATURE: _____ DATE: _____
WITNESSED BY: _____ DATE: _____

BOOK NO.

CONTINUED from page

82

CONTINUED on page

SIGNATURE: _____ DATE: _____
WITNESSED BY: _____ DATE: _____

BOOK NO.

CONTINUED from page

CONTINUED on page

SIGNATURE: _____ DATE: _____
WITNESSED BY: _____ DATE: _____

BOOK NO.

83

CONTINUED from page

84

CONTINUED on page

SIGNATURE: _____ DATE: _____
WITNESSED BY: _____ DATE: _____

BOOK NO.

CONTINUED from page

CONTINUED on page

SIGNATURE: _____ DATE: _____

WITNESSED BY: _____ DATE: _____

BOOK NO.

85

CONTINUED from page

CONTINUED on page

SIGNATURE: _____ DATE: _____
WITNESSED BY: _____ DATE: _____

BOOK NO.

CONTINUED from page

CONTINUED on page

SIGNATURE: _____ DATE: _____
WITNESSED BY: _____ DATE: _____

BOOK NO.

87

CONTINUED from page

CONTINUED on page

SIGNATURE: _____
WITNESSED BY: _____
DATE: _____
DATE: _____

BOOK NO.

CONTINUED from page

89

CONTINUED on page

SIGNATURE: _____ DATE: _____
WITNESSED BY: _____ DATE: _____

BOOK NO.

CONTINUED from page

92

CONTINUED on page

SIGNATURE: _____
WITNESSED BY: _____
DATE: _____
DATE: _____

BOOK NO.

CONTINUED from page

93

CONTINUED on page

SIGNATURE: _____ DATE: _____

WITNESSED BY: _____ DATE: _____

BOOK NO.

CONTINUED from page

CONTINUED on page

SIGNATURE: _____ DATE: _____
WITNESSED BY: _____ DATE: _____

BOOK NO.

95

CONTINUED from page

CONTINUED on page

SIGNATURE : _____ DATE : _____

WITNESSED BY : _____ DATE : _____

BOOK NO.

CONTINUED from page

97

CONTINUED on page

SIGNATURE: _____ DATE: _____
WITNESSED BY: _____ DATE: _____

BOOK NO.

CONTINUED from page

99

CONTINUED on page

SIGNATURE: _____ DATE: _____
WITNESSED BY: _____ DATE: _____

BOOK NO.

CONTINUED from page

100

CONTINUED on page

SIGNATURE: _____ DATE: _____
WITNESSED BY: _____ DATE: _____

BOOK NO.

CONTINUED from page

101

CONTINUED on page

SIGNATURE: _____
WITNESSED BY: _____
DATE: _____
DATE: _____

BOOK NO.

CONTINUED from page

102

CONTINUED on page

SIGNATURE: _____
WITNESSED BY: _____
DATE: _____
DATE: _____

BOOK NO.

CONTINUED from page

103

CONTINUED on page

SIGNATURE: _____
WITNESSED BY: _____
DATE: _____
DATE: _____

BOOK NO.

CONTINUED from page

105

CONTINUED on page

SIGNATURE: _____ DATE: _____
WITNESSED BY: _____ DATE: _____

BOOK NO.

CONTINUED from page

107

CONTINUED on page

SIGNATURE: _____
WITNESSED BY: _____
DATE: _____
DATE: _____

BOOK NO.

CONTINUED from page

108

CONTINUED on page

SIGNATURE: _____ DATE: _____
WITNESSED BY: _____ DATE: _____

BOOK NO.

CONTINUED from page

109

CONTINUED on page

SIGNATURE: _____ DATE: _____
WITNESSED BY: _____ DATE: _____

BOOK NO.

CONTINUED from page

111

CONTINUED on page

SIGNATURE: _____
WITNESSED BY: _____
DATE: _____
DATE: _____

BOOK NO.

CONTINUED from page

112

CONTINUED on page

SIGNATURE: _____ DATE: _____
WITNESSED BY: _____ DATE: _____

BOOK NO.

CONTINUED from page

113

CONTINUED on page

SIGNATURE: _____ DATE: _____

WITNESSED BY: _____ DATE: _____

BOOK NO.

CONTINUED from page

114

CONTINUED on page

SIGNATURE: _____ DATE: _____
WITNESSED BY: _____ DATE: _____

BOOK NO.

CONTINUED from page

115

CONTINUED on page

SIGNATURE: _____
WITNESSED BY: _____
DATE: _____
DATE: _____

BOOK NO.

CONTINUED from page

116

CONTINUED on page

SIGNATURE: _____ DATE: _____
WITNESSED BY: _____ DATE: _____

BOOK NO.

CONTINUED from page

117

CONTINUED on page

SIGNATURE: _____ DATE: _____
WITNESSED BY: _____ DATE: _____

BOOK NO.

CONTINUED from page

118

CONTINUED on page

SIGNATURE: _____ DATE: _____
WITNESSED BY: _____ DATE: _____

BOOK NO.

CONTINUED from page

119

CONTINUED on page

SIGNATURE: _____
WITNESSED BY: _____
DATE: _____
DATE: _____

BOOK NO.

CONTINUED from page

120

CONTINUED on page

SIGNATURE: _____ DATE: _____
WITNESSED BY: _____ DATE: _____

BOOK NO.

CONTINUED from page

122

CONTINUED on page

SIGNATURE: _____ DATE: _____
WITNESSED BY: _____ DATE: _____

BOOK NO.

CONTINUED from page

123

CONTINUED on page

SIGNATURE: _____
WITNESSED BY: _____
DATE: _____
DATE: _____

BOOK NO.

CONTINUED from page

124

CONTINUED on page

SIGNATURE: _____ DATE: _____
WITNESSED BY: _____ DATE: _____

BOOK NO.

CONTINUED from page

CONTINUED on page

SIGNATURE : _____ DATE : _____

WITNESSED BY : _____ DATE : _____

BOOK NO.

CONTINUED from page

127

CONTINUED on page

SIGNATURE: _____ DATE: _____
WITNESSED BY: _____ DATE: _____

BOOK NO.

CONTINUED from page

129

CONTINUED on page

SIGNATURE: _____ DATE: _____
WITNESSED BY: _____ DATE: _____

BOOK NO.

CONTINUED from page

130

CONTINUED on page

SIGNATURE: _____
WITNESSED BY: _____
DATE: _____
DATE: _____

BOOK NO.

CONTINUED from page

131

CONTINUED on page

SIGNATURE: _____ DATE: _____
WITNESSED BY: _____ DATE: _____

BOOK NO.

132

CONTINUED from page

133

CONTINUED on page

SIGNATURE: _____
WITNESSED BY: _____
DATE: _____
DATE: _____

BOOK NO.

CONTINUED from page

134

CONTINUED on page

SIGNATURE: _____ DATE: _____
WITNESSED BY: _____ DATE: _____

BOOK NO.

CONTINUED from page

135

CONTINUED on page

SIGNATURE: _____ DATE: _____
WITNESSED BY: _____ DATE: _____

BOOK NO.

CONTINUED from page

137

CONTINUED on page

SIGNATURE: _____
WITNESSED BY: _____
DATE: _____
DATE: _____

BOOK NO.

CONTINUED from page

138

CONTINUED on page

SIGNATURE: _____ DATE: _____
WITNESSED BY: _____ DATE: _____

BOOK NO.

CONTINUED from page

140

CONTINUED on page

SIGNATURE: _____ DATE: _____
WITNESSED BY: _____ DATE: _____

BOOK NO.

CONTINUED from page

142

CONTINUED on page

SIGNATURE:
WITNESSED BY:
DATE:
DATE:

BOOK NO.

CONTINUED from page

143

CONTINUED on page

SIGNATURE: _____ DATE: _____

WITNESSED BY: _____ DATE: _____

BOOK NO.

CONTINUED from page

144

CONTINUED on page

SIGNATURE: _____
WITNESSED BY: _____
DATE: _____
DATE: _____

BOOK NO.

CONTINUED from page

145

CONTINUED on page

SIGNATURE : _____ DATE : _____
WITNESSED BY : _____ DATE : _____

BOOK NO.

CONTINUED from page

CONTINUED on page

SIGNATURE: _____ DATE: _____
WITNESSED BY: _____ DATE: _____

BOOK NO.

146

CONTINUED from page

148

CONTINUED on page

SIGNATURE: _____ DATE: _____
WITNESSED BY: _____ DATE: _____

BOOK NO.

CONTINUED from page

149

CONTINUED on page

SIGNATURE: _____ DATE: _____
WITNESSED BY: _____ DATE: _____

BOOK NO.

CONTINUED from page

150

CONTINUED on page

SIGNATURE: _____ DATE: _____
WITNESSED BY: _____ DATE: _____

BOOK NO.

CONTINUED from page

151

CONTINUED on page

SIGNATURE: _____ DATE: _____
WITNESSED BY: _____ DATE: _____

BOOK NO.

CONTINUED from page

152

CONTINUED on page

SIGNATURE: _____
WITNESSED BY: _____
DATE: _____
DATE: _____

BOOK NO.

CONTINUED from page

153

CONTINUED on page

SIGNATURE: _____ DATE: _____
WITNESSED BY: _____ DATE: _____

BOOK NO.

CONTINUED from page

154

CONTINUED on page

SIGNATURE: _____ DATE: _____
WITNESSED BY: _____ DATE: _____

BOOK NO.

CONTINUED from page

155

CONTINUED on page

SIGNATURE: _____ DATE: _____

WITNESSED BY: _____ DATE: _____

BOOK NO.

CONTINUED from page

156

CONTINUED on page

SIGNATURE: _____ DATE: _____
WITNESSED BY: _____ DATE: _____

BOOK NO.

CONTINUED from page

157

CONTINUED on page

SIGNATURE: _____ DATE: _____
WITNESSED BY: _____ DATE: _____

BOOK NO.

CONTINUED from page

159

CONTINUED on page

SIGNATURE: _____ DATE: _____

WITNESSED BY: _____ DATE: _____

BOOK NO.

CONTINUED from page

160

CONTINUED on page

SIGNATURE: _____ DATE: _____
WITNESSED BY: _____ DATE: _____

BOOK NO.

CONTINUED from page

161

CONTINUED on page

SIGNATURE: _____ DATE: _____

WITNESSED BY: _____ DATE: _____

BOOK NO.

CONTINUED from page

162

CONTINUED on page

SIGNATURE: _____ DATE: _____
WITNESSED BY: _____ DATE: _____

BOOK NO.

CONTINUED from page

163

CONTINUED on page

SIGNATURE: _____ DATE: _____
WITNESSED BY: _____ DATE: _____

BOOK NO.

CONTINUED from page

164

CONTINUED on page

SIGNATURE: _____ DATE: _____

WITNESSED BY: _____ DATE: _____

BOOK NO.

CONTINUED from page

165

CONTINUED on page

SIGNATURE: _____ DATE: _____
WITNESSED BY: _____ DATE: _____

BOOK NO.

CONTINUED from page

167

CONTINUED on page

SIGNATURE: _____
WITNESSED BY: _____
DATE: _____
DATE: _____

BOOK NO.

CONTINUED from page

168

CONTINUED on page

SIGNATURE: _____ DATE: _____
WITNESSED BY: _____ DATE: _____

BOOK NO.

CONTINUED from page

169

CONTINUED on page

SIGNATURE: _____ DATE: _____

WITNESSED BY: _____ DATE: _____

BOOK NO.

CONTINUED from page

170

CONTINUED on page

SIGNATURE: _____
WITNESSED BY: _____
DATE: _____
DATE: _____

BOOK NO.

CONTINUED from page

172

CONTINUED on page

SIGNATURE: _____ DATE: _____
WITNESSED BY: _____ DATE: _____

BOOK NO.

CONTINUED from page

173

CONTINUED on page

SIGNATURE: _____ DATE: _____
WITNESSED BY: _____ DATE: _____

BOOK NO.

CONTINUED from page

174

CONTINUED on page

SIGNATURE: _____ DATE: _____
WITNESSED BY: _____ DATE: _____

BOOK NO.

CONTINUED from page

175

CONTINUED on page

SIGNATURE: _____ DATE: _____
WITNESSED BY: _____ DATE: _____

BOOK NO.

CONTINUED from page

176

CONTINUED on page

SIGNATURE: _____
WITNESSED BY: _____
DATE: _____
DATE: _____

BOOK NO.

CONTINUED from page

177

CONTINUED on page

SIGNATURE: _____ DATE: _____
WITNESSED BY: _____ DATE: _____

BOOK NO.

CONTINUED from page

178

CONTINUED on page

SIGNATURE: _____ DATE: _____
WITNESSED BY: _____ DATE: _____

BOOK NO.

CONTINUED from page

179

CONTINUED on page

SIGNATURE: _____
WITNESSED BY: _____
DATE: _____
DATE: _____

BOOK NO.

180

CONTINUED from page

181

CONTINUED on page

SIGNATURE: _____
WITNESSED BY: _____
DATE: _____
DATE: _____

BOOK NO.

CONTINUED from page

182

CONTINUED on page

SIGNATURE: _____ DATE: _____
WITNESSED BY: _____ DATE: _____

BOOK NO.

CONTINUED from page

183

CONTINUED on page

SIGNATURE: _____
WITNESSED BY: _____
DATE: _____
DATE: _____

BOOK NO.

CONTINUED from page

184

CONTINUED on page

SIGNATURE: _____ DATE: _____
WITNESSED BY: _____ DATE: _____

BOOK NO.

CONTINUED from page

185

CONTINUED on page

SIGNATURE: _____ DATE: _____
WITNESSED BY: _____ DATE: _____

BOOK NO.

CONTINUED from page

186

CONTINUED on page

SIGNATURE: _____
WITNESSED BY: _____
DATE: _____
DATE: _____

BOOK NO.

CONTINUED from page

187

CONTINUED on page

SIGNATURE: _____ DATE: _____
WITNESSED BY: _____ DATE: _____

BOOK NO.

CONTINUED from page

188

CONTINUED on page

SIGNATURE: _____ DATE: _____
WITNESSED BY: _____ DATE: _____

BOOK NO.

CONTINUED from page

189

CONTINUED on page

SIGNATURE: _____ DATE: _____
WITNESSED BY: _____ DATE: _____

BOOK NO.

CONTINUED from page

190

CONTINUED on page

SIGNATURE: _____
WITNESSED BY: _____
DATE: _____
DATE: _____

BOOK NO.

CONTINUED from page

191

CONTINUED on page

SIGNATURE: _____ DATE: _____

WITNESSED BY: _____ DATE: _____

BOOK NO.

CONTINUED from page

192

CONTINUED on page

SIGNATURE: _____ DATE: _____

WITNESSED BY: _____ DATE: _____

BOOK NO.

CONTINUED from page

194

CONTINUED on page

SIGNATURE : _____ DATE : _____
WITNESSED BY : _____ DATE : _____

BOOK NO.

CONTINUED from page

195

CONTINUED on page

SIGNATURE: _____ DATE: _____
WITNESSED BY: _____ DATE: _____

BOOK NO.

CONTINUED from page

196

CONTINUED on page

SIGNATURE: _____ DATE: _____
WITNESSED BY: _____ DATE: _____

BOOK NO.

CONTINUED from page

197

CONTINUED on page

SIGNATURE: _____ DATE: _____
WITNESSED BY: _____ DATE: _____

BOOK NO.

CONTINUED from page

198

CONTINUED on page

SIGNATURE: _____ DATE: _____
WITNESSED BY: _____ DATE: _____

BOOK NO.

CONTINUED from page

199

CONTINUED on page

SIGNATURE: _____
WITNESSED BY: _____
DATE: _____
DATE: _____

BOOK NO.

CONTINUED from page

200

CONTINUED on page

SIGNATURE: _____ DATE: _____
WITNESSED BY: _____ DATE: _____

BOOK NO.

Made in the USA
Monee, IL
22 January 2021